GUIDE for THE PROCUREMENT of ENGINEERED EQUIPMENT

A Simplified 12-Step Procurement Process

By Ramesh (RD) Patel

ISBN: 1481944894

ISBN 13: 9781481944892

Contents

List of Tables

Preface

This guide provides guidance for the procurement of engineered equipment for medium- to large-size projects and is based on my personal experience after more than forty years in engineering and management in the nuclear power plant industry. During that time, I have been given the task of leading engineering, procurement, and quality assurance coordination for the successful procurement of all new nuclear power plant equipment.

I struggled at first, and could not find any useful and available procurement/process guides or resources. So, I started with the fundamentals and formed a practical solution with all the team members (engineering, procurement, and quality assurance) involved to develop an efficient and effective procurement process for the project, which was proved successful in delivering over one billion dollars of equipment for a new nuclear power plant project.

This guide covers the details and provides planning guidance for all phases of successful procurement in just twelve steps, from writing a technical specification, to the receipt inspection, and finally to the acceptance of equipment at the final destination, the jobsite.

My past struggles inspired me to document all that I've learned for future users who are currently involved in any phase of procuring engineered equipment and those who may be involved in any large, medium, or small size project, such as a power plant (nuclear, solar park, coal, or gas type), refinery, small chemical/industrial plant, or similar project in the future. Certain elements of this guide can be effectively applied to a small-scale project, such as a home building or modification project, where a number of engineered equipment pieces and major home appliances or systems must be procured or installed.

The guide also can be used for training the assigned engineers, as well as the procurement and quality assurance team members, vendors, or even for self-education.

RAMESH (RD) PATEL

Acknowledgements

The author is grateful to Samir Riad for inspiring him to write this book and for giving him the initial guidance.

The author appreciates Richard Wolters and David Faulstich for their valuable contributions to this guide.

Finally, thanks to the CreateSpace publishing team for their editing and layout support to make this book presentable.

Chapter 1
Introduction

Procurement is the process of procuring or purchasing, sometimes called *buying*. The term buying is most familiar to all, since everyone buys something every day. Everyone has had the experience, at one time or another, of buying items such as groceries, clothing, appliances, automobiles, etc. To buy everyday items requires some effort in the selection, shopping around for the best price, availability in the vicinity, and, of course, the quality. Sometimes, quality can be subjective rather than objective, meaning that it can be based on personal choice or preference of an item.

How often have you learned that the so-and-so project has increased in cost by three to four or even five times the original estimate, and its completion timeline has been delayed or extended significantly? Similarly, many people may have experienced higher costs and time delays in their own personal projects, such as home modification or getting the delivery of an item ordered. Most of the cost and schedule delays are attributed to scope or design changes after the contract is awarded, and in some cases, work may have already begun. Another major cause could be a change in regulations. For example, in a home modification project, city ordinances that may have been changed since the home was built, can impact the project's cost and schedule. The impact due to such rule changes may not be known until the work has begun. At times, even the contractors could not have known until they got involved in the work.

Therefore, the need for carefully documenting the initial requirements, and especially the changes and their attendant cost and schedules, is crucial. Also, it is important to have available a review and approval process based on the impact, to allow small decisions to be made by the responsible individual while delegating larger impacts to management.

When engineered items, such as equipment or a subsystem for a project, are needed, then a significant effort is required in designing, creating technical specifications, shopping around for a vendor (sometimes called a *supplier*, who can provide the best price and quality for a safe and efficient operation), as well as maintenance and repair support, during the intended life of the equipment operation.

To help current and future engineers, and others involved in designing and procuring engineered equipment, as well as those users for personal projects, this guide attempts to capture and document pertinent information based on my forty years of experience in the nuclear industry, including the development of procurement processes for engineered equipment.

All the examples used in this guide are based on my experience in the procurement of engineered equipment for nuclear power plant applications. Therefore, any requirements, codes, or regulations cited and examples included in this guide are applicable to nuclear power plants. However, the processes and guidance are equally applicable to other industries, because the nuclear industry has stricter quality and safety rules than others, and it is heavily regulated by the United States Nuclear Regulatory Commission (USNRC) and/ or by similar authorities overseas. It should be noted that most of the overseas authorities have applied or, in some cases, developed similar rules to USNRC rules for their respective country's nuclear power plant regulations.

Chapter 2
Overview

A complete procurement process requires three responsible parties: 1) responsible engineer, 2) quality assurance (QA) engineer, and 3) procurement agent, also called the buyer. One of the above responsible parties may take the lead responsibility for a particular phase, with significant support from the other two. Lead responsibility of a phase may be assigned based on the party that is required to provide majority input to that phase.

The guidance is divided into three phases. **Phase I:** development of a technical specification through signing a contract with the selected vendor. **Phase II:** shop activities, such as fabrication, testing, inspections, and release for shipment. **Phase III:** shipping from the vendor shop to delivery at the jobsite and receipt inspection follow-up with the customer's final acceptance.

Phase I covers development of detail requirements. It starts with the review and incorporation of customer requirements, both technical and administrative, into documents and procedures of both. A typical equipment specification document requires functional, safety-related equipment qualifications, testing, and other related codes and standards. Along with a specification, equipment details and technical data must be provided in one or more of these three formats: 1) detail drawings, specifically for built-to-print, custom-designed equipment; 2) data sheets with technical details for equipment potentially available from vendors or that may require minor modifications, e.g., valves, pumps, heat exchangers, etc.; and 3) design/data lists (directly from the design database) for the commodity type of equipment, e.g., pipes, general purpose valves, pipe supports, cables, instruments, etc.

With technical specifications, the request for proposal (RFP) must include documents covering commercial requirements, administrative requirements, and quality assurance requirements. The next step is obtain proposals, also called *bids*, from qualified vendors, and selects the vendor using the established selection criteria. Finally, sign the contract with the selected vendor. Use sourcing to search for qualified potential vendors for the type of equipment to be procured.

Phase II covers a preproduction meeting and review of the detailed documents submittal list by the vendor. The vendor is to develop a fabrication plan, a detailed test plan, and a detailed list of inspection, witness, and hold points. The documents identified as required before fabrication/production starts must be reviewed and approved prior to this process. If any documents are rejected for any reason during the review, those documents must be resubmitted with resolution of all comments, and final approval must be obtained prior to the start of fabrication/production. Some of the documents require review/approval prior to supplier design, manufacturing, testing, or delivery, as identified on the vendor's submittal list.

The last step in phase II is to perform a final inspection, including source inspection for material traceability and approval of quality records, prior to release of the equipment for shipment or delivery to the jobsite.

The final phase, phase III, covers the shipping process to be used to deliver the equipment to the jobsite via a specified transportation method. Upon the arrival of the equipment at the jobsite, a receipt inspection shall be performed by the onsite quality assurance team. Any discrepancies from this receipt inspection should be noted or documented, and all issues must be resolved or rectified by the vendor, to the satisfaction of the customer, prior to final acceptance.

The procurement cycle timeline depends on the type and complexity of the equipment or subsystem. Lead times for equipment should be established based on the procurement cycle. The lead time should be coordinated with the need of equipment at the jobsite for installation in the project construction plan/schedule (e.g., when building a home, make sure that all major appliances, equipment, and materials are procured in a timely manner in order to support the construction sequence and schedule). A typical overall procurement process is summarized in table 1.

Table 1 – A Typical Procurement Process Overview

Phase I RFP, evaluation of bids, and contract	Phase II Fabrication, tests, and inspections	Phase III Shipping, receipt inspection, and final acceptance
Issue purchase specification package	Preproduction review	Shipment transported to the jobsite
Administrative and commercial requirements	Review vendor quality and fabrication plans	Arrived at the site and stored in a proper storage
Quality assurance (QA) requirements	Review vendor test and inspection plans, including witness and hold points	Site QA perform receipt inspection and document findings, if any
Qualify potential vendors for QA and commercial acceptance	Fabrication, in-process inspections, and witness key tests	Resolution of findings with vendor
Issue request for proposal and obtain bids or proposals from qualified vendors	Review and acceptance of quality records, equipment inspection, and final quality certification	Upon resolution of findings, bring equipment to its original condition, as applicable
Evaluate bids and select vendor for final negotiations	Final review and acceptance of equipment shipment	**Approval of final acceptance**
Sign the contract	**Release of shipment to jobsite**	

Chapter 3
Phase I: From Technical Specification to Signing a Contract with a Vendor

This phase requires engineering to prepare a quality assurance requirements document and technical specifications based on the systems and/or plant design requirements, including detailed drawings, data sheets, and/or a data list that can be generated directly from the design database (see table 2 example). To avoid costly delays, all the design changes related to the subject equipment specification must be incorporated prior to signing a contract with the selected vendor. Any subsequent design changes must be controlled, and all the alternate technical options need to be evaluated before making any further changes to the signed contract. No more technical changes are to be accepted after the fabrication begins, in order to avoid significant cost and schedule impacts, unless it is cost-beneficial to the overall project.

Table 2 – Example of a Data List – General Services Valves

Procurement Specification Number and Title: _____

Revision No._____ , Date:_____

Component Identification Number	Valve – 1000	Valve – 1002	Valve – 1003
Abbreviated Description	Containment Isolation Valve		
Valve Type / (End-to-End Dimensions for Large Valves, i.e., 300 mm and over)	Globe Valve		
Inlet size/Outlet size	80 mm/80 mm		
Valve Class/Body Material	CL 1500/316L Stainless Steel		
End Prep for Inlet/Outlet	Butt Weld 80S		
Design Pressure/Design Temperature	8600 kpa/300 degree C		
Safety/Design Code	S-Safety/ASME Code Class 1		
Quality Class/Seismic/ Class1E	S-safety/ Seismic Category 1/1E		
Building/Room Number	Reactor Building/ Room 100		
Remarks			

Step 1: The Technical and Purchase Specification

A typical purchase specification addresses the following elements:

1. **Scope:** The scope describes what is covered in this specification. It also identifies the equipment, including a description, identification number, and quantity. The specification must address the entire set of requirements related to design, fabrication, manufacturing, testing, packaging, and shipping or delivery of the equipment, with complete accessories, appurtenances, and spare parts.

2. **Codes and Standards:** List all applicable industry and regulatory codes and standards in this section. Examples include industrial codes and standards, such as the American Society of Mechanical Engineers (ASME), Institute of Electrical and Electronics Engineers (IEEE), etc., and for regulatory standards, such as the United States (US) Code of Federal Regulations (CFR), 10CFR50, and US Regulatory Guides (RG) for nuclear power plants application.

3. **Description:** This section includes physical and functional descriptions of the equipment. The physical description can be illustrated in a

drawing with proper orientation and required service points, such as air, water, or power connections. The functional information should be limited to only those functions the supplier needs to know for proper manufacturing and testing of the equipment.

4. **Design Requirements**: This section includes equipment classification. For example, for nuclear power plant equipment, the classification shall include quality (nuclear quality as defined in 10CFR50, appendix B), safety, seismic, and applicable code classes, such as ASME Section III, Class 1, 2, or 3.

5. **Design Life**: Equipment design life. The typical nuclear power plant design life is forty to sixty years. Equipment should be qualified for design and operation life for the given environmental conditions. The equipment's qualified life may be shorter than the plant design life. In that case, replacement schedule to be established in the plant maintenance program.

6. **Design and Operating Conditions**: Design conditions and operating conditions for each mode of operation to be provided here, or can reference the appropriate data sheet, data list, or specific drawing.

7. **Equipment Qualification**: Define all loads, load combinations, and conditions (seismic, dynamic, and environmental) with acceptance criteria, or reference the applicable codes and standards for acceptance criteria of equipment qualification. Example: Nuclear power plant equipment application and environmental, seismic, and dynamic conditions typically are provided in an appendix, on a data sheet, or in a specific document or drawings. The loads and load combinations are provided in a similar manner for final equipment qualification.

8. **Combustible Load Requirement**: This section includes information for the vendor to be able to determine and provide the combustible load associated with the scope of this equipment, including spare parts. Examples of combustible loads are: liquids such as fuels, oils, and lubricant; electrical insulation such as cable insulation, motor winding insulation, and thermal insulation with any plastic parts; coatings such as paint; and other items such as seals and gaskets. Typical format of supplier data should be included for each piece of equipment with its material specification, quantity of material in kilograms or liters, or the appropriate unit to be specified here. This is required to design and develop a fire

protection program in a nuclear power plant facility per 10CFR50, appendix R.

9. **Noncombustible Material**: Example: for nuclear application requirements as defined by National Fire Protection Association (NFPA) 804, Standard for Fire Protection of Advanced Light Water Reactor Electric Generating Plants. A similar applicable standard to be referenced here for other industry applications.

10. **Special Conditions**: Define any other special conditions of this equipment's design, material, or services, if applicable.

11. **Fabrication Requirements**: Add any specific fabrication requirements for the type of equipment being furnished or purchased, or the type of service being performed. Any special welding (e.g., inertia welding) or other specific processes can be included in this section. These can be provided in a separate drawing or data sheet.

12. **Materials**: Provide material requirements here, or refer to any applicable generic material specifications for the project. Define material for major components and for subcomponents of this equipment. Example: for a nuclear application, ASME materials to be used for major components and American

Society of Testing and Materials (ASTM) materials to be used for subcomponents. Also, appropriate material to be specified for non-metallic components, such as seals, gaskets, paint, etc.

13. **Inspection and Testing**: Define applicable codes and standards for equipment inspection and testing here, or refer to the appropriate project document.

   a. Example used for nuclear application: The purchaser or representative reserves the right of twenty-four-hour access to the vendor shop (including the vendor's subcontractors' shops) while equipment is being fabricated or manufactured. Also, the purchaser may request notification from the supplier for intermediate inspections of fabricated work. The vendor's responsibilities include conducting all examinations, inspections, and tests (including recording of results and maintaining records thereof), and furnishing all required testing materials and equipment, including required certifications, if any.

   b. The vendor shall provide access to the purchaser or representative, upon request, at any time, of related vendor documents for fabrication and testing, including the vendor's subcontractors' documents.

c. Define preliminary inspection and testing witness and hold points here or in a separate attachment to this RFP package for bidding purposes. Also, after the contract, this list of information will help the vendor develop a detailed inspection and test plan with witness and hold points for purchaser's review and approval, and followup during the fabrication/manufacturing.

14. **Delivery Requirements**: This section includes the required delivery information for equipment identification and marking, nameplate, painting, coating, surface preparation, cleaning, packaging, and shipping. The vendor shall provide detailed storage and handling requirements for receiving equipment at the site. For example, in a nuclear project, generic requirements documents shall be prepared for each of the above areas that include customer's (plant owner's) specific requirements.

15. **Documents Submittal**: This section includes a preliminary documents submittal list for review and/or approval, to be defined here or in a separate list attached to the RFP package for bidding

purposes. After the contract is signed, the vendor will develop a detailed documents submittal list for final review and approval. The vendor will then submit all required documents prior to fabrication and obtain the purchaser's approval. Any special document or drawing required from the vendor for review or for information is to be included. If handover of the equipment to the customer will be delivery at the jobsite only—where installation, testing, and start-up responsibilities are with the customer—then a statement regarding special emphasis on related documents is needed to assure a clean handover. Examples: an equipment outline drawing showing all service-required interfaces and locations, a component information sheet, or an equipment spare parts list, including any special tools necessary for installation, operation, and maintenance of the scope of this equipment.

16. **List of Supporting Documents**: In this section, provide a list of all supporting documents to be provided by the buyer, including any attachments. The examples of key documents listed below are for reference only.

   a. Equipment layout drawing with physical envelop dimensions and services interfaces.

    b. Equipment qualification (EQ) requirements (for safety-related equipment as applicable to the nuclear industry). Other industries may have similar requirements.

    c. Cleaning, packaging, shipping, and storage and handling requirements.

    d. Quality assurance requirements.

    e. Weld preparation requirements and pipe size data.

    f. Electrical equipment general requirements, such as for wiring, terminal blocks, grounding, incoming voltage, etc.

    g. Painting, coating, and surface preparation requirements.

    h. Plant working fluids requirements.

Step 2: Administrative and Commercial Requirements

This section provides administrative requirements for the vendor or supplier, including specific requirements imposed by the customer. Examples of key areas to be addressed are listed below for reference.

1. Terms and conditions of the contract, including liability and indemnity clauses.

2. Detailed payment and warranty plan.

3. Responsibilities of vendor and vendor's sub-suppliers, if any.

4. Buyer review of procurement documents with the vendor.

5. Vendor support of audits and inspections.

6. Equipment and component identification and marking.

7. Vendor retention of quality records (e.g., for the life of the equipment).

8. Packaging and shipping.

9. Spare parts and special tools (e.g., vendor to supply all mandatory spare parts and special tools with the equipment).

10. All equipment manuals (e.g., quality, users, installation, etc.).

11. Delivery and schedule.

12. Pricing (e.g., fixed or unit price for commodities such as pipes, valves, standard instruments, etc., and how to handle any increase in scope or quantity for commodities).

13. Notifications and communications between the parties.

One of the critical functions in the commercial area is sourcing, which refers to a number of procurement practices aimed at finding, evaluating, and engaging suppliers of goods and services. Sourcing is used to identify potential vendors or suppliers, prior to sending the request for proposal packages. The following major factors shall be evaluated before prequalifying the vendors for a type of equipment or service.

—Financial capacity and stability
— Resource capability
— Shop and manufacturing capacity
— Track records, references, experience, etc.
— Quality, QA program, certifications, etc., which may require sending a quality team to audit vendor for prequalification
— Servicing capability throughout the life of equipment, especially for critical equipment

Step 3: Quality Assurance Requirements

Quality assurance requirements are critical for safety and compliance with the regulatory requirements of the applicable industry. For nuclear power plant applications, ASME/ANSI has developed the standard NQA-1 that documents requirements for quality assurance in nuclear facility applications. The standard provides eighteen basic requirements based on the criteria in appendix B to the code of federal regulations, title 10, part 50 (10CFR50). The NQA-1 standard also provides supplemental guidance for each criterion with additional non-mandatory guidance.

For nuclear power plant-related equipment with safety and ASME Section III code classes, vendors must have an NQA-1-compliant quality assurance program and must be audited for their qualification prior to requesting any bids from them. NQA-1-compliant vendors require an applicable ASME code certification. Vendors of non-safety or non-ASME types of equipment may need to be in compliance with the ISO 9001 quality management standard, which requires the vendor to obtain ISO 9001 certification. ISO is the International Organization for Standardization, and its standards are widely used in many industries internationally. Otherwise, the manufacturer's standards may apply to general-use type, or non-safety and non-ASME, equipment.

Key requirements, based on my nuclear experience, to be included in this quality assurance requirements document are listed below, for reference only.

1. Responsibilities of vendor and sub-vendor.
2. Quality assurance program (e.g., quality classification for safety and non-safety, applying NQA-1

and ISO 9001, respectively). See table 3 example of a typical quality assurance program structure and associated compliance standards.

3. Commercial grade dedication for using any dedicated component or part in safety class equipment. If vendor uses any commercial grade item in safety-related equipment for nuclear plants application, those items must be identified and qualified by using the established dedication process. A commercial grade item is defined by the NQA-1-1986 standard as follows:

   Commercial grade item is an item satisfying (a), (b), and (c) below:

   a.) *not subject to design or specification requirements that are unique to nuclear facilities;*

   b.) *used in applications other than nuclear facilities;*

   c.) *is to be ordered from the manufacturer/supplier on the basis of specifications set forth in the manufacturer's published product description (for example catalog).*

4. Inspection and test plans, including witness and hold points.

5. Documents submittal required by the buyer or purchaser.

6. Nonconformance issues and disposition of vendor deviations.

7. Submittal of a quality assurance manual and plans.

8. Vendor or supplier change request.

9. Surveillance and audits by buyer (see table 4).

10. Quality records required by buyer (see table 5).

11. Certification and release for shipment, including any applicable product quality certificates (see table 6).

Example forms with instructions for items 4, 5, 6, and 8 may be included in this quality assurance requirements document for the accurate tracking and recording of all such activities. The information provided in these example forms with the RFP will give initial input to vendors in the development of detailed plans, the change process, listings, and/or submittals after the contract is signed.

Table 3 – A Typical Quality Assurance Program Structure and Associated Compliance Standards

• 10CFR50, Appendix B – Quality assurance requirements for a nuclear facility (eighteen criteria)
  1. Organization
  2. Quality assurance program
  3. Design control
  4. Procurement document control
  5. Instructions, procedures, and drawings
  6. Document control
  7. Control of purchased items and services
  8. Identification and control of items
  9. Control of process
  10. Inspection
  11. Test control
  12. Control of measuring and test equipment
  13. Handling, storage, and shipping
  14. Inspection, test, and operating status
  15. Control of nonconforming items
  16. Corrective action
  17. Quality assurance records
  18. Audits
	• Implementing Standards
	  −ANSI 45.2 + daughter standards
	  −IEEE standards for Class 1E
	  −ASME NQA-1 standard with supplementary requirements and non-mandatory guidance
	  −NQA 2.7 (software)
	  −NRC regulatory guides
	  −Other codes and standards
	  −ASME code
	  −ISO 9001 standard
	  −10CFR21
	  −10CFR71
	• NRC generic communications
	  −Administrative letters
	  −Bulletins
	  −Circulars
	  −Generic letter
	  −Information notices
	  −Regulatory issue summaries
	• NRC and customer audits
	• Customer requirements

Table 4 – Example of Generic Audit Checklist for Quality

Supplier Information: Name, Address, Phone Number, and QA Management Contact:_____ ASME Code Certification Number(s) and Expiration Dates: _____ ISO Certification Number(s) and Expiration Dates: _____	

Audit Number and Date(s): _____
Audit Team: _____
Supplier QA Manual Number, Revision, and Date: _____

Audit Scope: *Any one or a combination of the following standards or requirements:*
ASME NQA-1, ASME Section III, ISO 9001, ASME NQA-2A, PART 2.7, ISO 90003, Project QA Requirements Document
Product Services Class: *Safety or non-safety or generic*

Audit Section	Section Description	Implementation Status	Comments/ Findings
1	Program compliance		
2	Organization/Program		
3	Nonconforming items/10CFR21		
4	Audits		
5	Corrective action		
6	Training/Certification		
7	Records		
8	Design		
9	Procurement & source material		
10	Document control		
11	Material control, handling, shipping, and storage		
12	Fabrication, assembly, and special processes		
13	Inspection and test		
14	Calibration		
15	Commercial grade dedication		
16	Software quality assurance		

Implementation Key: S = Satisfactory, U = Unsatisfactory, N/A = Not Applicable

This checklist is to be used as a guideline in conjunction with specific requirements of the appropriate industry documents imposed by the procurement documents.

Table 5 – Example of Typical Quality Records Outline

Equipment Package Name and Number:
1. *Material*
2. *Qualification records*
3. *Welding*
4. *Nondestructive testing (NDT)*
5. *Post-weld heat treatment*
6. *Final fabrication, inspection, and test records*
7. *Document control records*
8. *Radiographs*
9. *Deviations*
10. *Environmental qualification records*
11. *Special notes (if any)*

Table 6 – Example of Typical Quality Certificate

Equipment Package Contract Number:	
Equipment Package Number and Title:	

Vendor Certification:
This is to certify that the equipment identified herein has been manufactured under a controlled quality assurance program and is in conformance with the procurement quality requirements, including applicable codes, standards, and specifications as identified in the above contract documents, unless noted below. Any supporting documentation will be forwarded or retained in accordance with the contract requirements.

Signed:_____ Date:_____ Title:_____

Buyer's Quality Assurance Certification:
The equipment described above was procured in accordance with the above contract quality assurance program (list any document number with the final revision number). This is to certify that evidence supporting the above vendor's certification statement has been reviewed and no equipment nonconformance (any deviations) from procurement quality requirements has been found unless noted below.

Signed:_____ Date:_____ Title:_____

Deviations and Closure:
List any deviation by number and title with its closure/approval date.

Equipment Serial Numbers:	Vendor to provide.
Nuclear (or applicable) Safety Related?	Yes or No
ASME Code:	Yes or No
Quality Class:	S for Safety or G for Generic (non-safety)

Buyer or Customer Representative:
The shipment has been reviewed for documents completeness and the equipment, as identified in this quality certificate, is released for shipment.
Signed:_____ Date:_____ Title:_____

Step 4: Regulatory Requirements

This section provides any regulatory requirements applicable to the industry where equipment will be used. Many governments have endorsed and accepted as safety codes for a type of industry, such as ASME codes for pressure boundary components, NFPA codes for fire protection equipment, and API codes for equipment used in the oil and gas industry, etc.

Nuclear example: For any safety-related designation with quality class "S" items, the vendor or supplier shall be responsible for the reporting of significant defects and deficiencies as defined in the United States Code of Federal Regulations, title 10, part 21 (10CFR21) to the purchaser or buyer.

Step 5: Request for Proposal, Bid Evaluation, and Signing a Contract

This process follows the generation of the request for proposal (RFP) package using the documents discussed previously. Then the RFP packages are sent to all qualified potential vendors for submitting their respective bids. After the bids are received, evaluate them, select a bidder, and sign a contract with or issue a purchase order to the selected vendor. Each of these elements is reviewed in detail below.

1. **Request for Proposal**: The request for proposal package shall include the following documents:

   a. An invitation to submit a proposal or bid (the RFP), instruction to bidders, bidding requirements, and proposal data form (see table 7 example).

   b. Administrative and commercial requirements pertaining to procurement from the customer and the purchaser's specific requirements. This includes a vendor documents submittal list with a schedule.

   c. Purchase and technical specifications with all supporting documents and applicable data sheets, equipment details data list (see table 2), and/or specific drawings.

   d. A quality assurance requirements document. This should include a required inspection and testing notification list, including key witness and hold points, a quality records outline, and a documents submittal list. After signing a contract, the vendor shall develop the above items in detail and submit them with a quality plan for buyer's review and approval, to be used in the subsequent tracking of progress until delivery.

Table 7 – Generic Proposal Data Request Example[1]

Typical information to be requested with the RFP for bidders to submit with their respective proposal, used for evaluating the bids.

Equipment Procurement Package Number & Name:	Bidder's Name
Outline Drawing with Overall/Critical Dimensions	
Weight (Net, Shipping, etc.)	
Service Interfaces:	
Nozzle Load Allowable	
Power Requirements (Voltage, etc.)	
Water, Air, etc.	
Example for a Typical Heat Exchanger:	
Total effective tube surface, *sq. meter*	
Number of tubes required	
Length of tubes, *meter*	
Effective length	
Total length	
Tube manufacturer	
Tube material (by common name and ASTM number)	
Tube sheet material	
Unit weight of tubes, *kg per meter*	
Total weight of tubes, *kg*	

Outside diameter of tubes, *mm*	_____
Maximum	
Minimum	
Wall thickness of tubes, *mm*	_____
Maximum	_____
Minimum	

(1) *For safety class equipment, as applicable, bidders are to provide component, parts, and subassembly level classification plans with appropriate justification and equipment qualification (EQ) plan, including the method of qualification, with their respective bid.*

2. **Evaluation of Bids:** Bids from all vendors to be tabulated for comparison and ultimate selection. Table 8 provides example of key elements for consideration in the bids evaluation and risk assessment. The evaluation includes but is not limited to the following:

a. Technical exception to be identified and valued accordingly. Examples: different materials proposed, which may not be acceptable, different instrumentation plan proposed, low-allowable nozzle loads proposed, etc.

b. Design impact due to difference proposed (e.g., physical layout, piping and/or instrumentation design, service and maintenance requirements, etc.).

c. Commercial exceptions, such as delivery date, method of transportation for shipment, warranty, etc.

d. Other considerations, such as past work experiences. Examples: 1) for deviations, repairs, meeting schedule, etc.; 2) equipment quality and track record; and 3) licensing commitment. Also, any special requirements or options each vendor may have provided in its proposal for evaluation.

e. Vendor's quality assurance manual has been reviewed and is acceptable or any exceptions noted, etc.

Table 8 – Example of Key Elements for Consideration in the Bids Evaluation and Risk Assessment

1	Review each submitted proposal for completeness and request missing information or data from vendors accordingly.
2	Identify vendors who are not providing a complete scope of supply. Contact those vendors for clarification as needed.
3	Determine if a partial scope of supply proposal is acceptable and review with affected technical teams. If not acceptable, consider excluding affected vendor from the bid evaluation; and document information in the bid evaluation record. If acceptable, contact other vendors and obtain adjustment in proposal price.
4	Confirm each proposal meets target price. Contact high price (i.e., over the target price) vendors, as needed, to discuss possible cost reduction in order to avoid the re-bidding.
5	Review the technical and commercial exceptions taken to the RFP in each proposal.
6	Review proposals for the technical and commercial adequacy and validity or acceptance of the exceptions taken by each vendor.
7	Document all exceptions taken to both the technical and commercial portions of the RFP by each vendor providing an acceptable proposal.
8	Determine lowest evaluated bidder based on the exceptions taken to the RFP by each bidder and associated adjustment of cost (i.e., adders and/or subtractions to the bid value). Adders may include such items as changes to system physical layouts, impact on system operation, needed special spare parts, etc. Subtractions may include items such as, providing a higher quality standard product, standard operational accessories not required by the base RFP, etc.

3. **Signing a Contract:** All exceptions found in the bid evaluations must be resolved with the vendors under consideration. Consider the adjustment of each resolution in comparison with all the bidders. Based on the comparison, recommend vendor to be selected for this equipment package. Notify the selected vendor for final negotiations/discussion and then sign the contract.

Chapter 4
Phase II: Preproduction, Fabrication, Testing, Inspections, and Release for Shipment

This phase requires quality assurance and engineering to review, monitor, and track all the activities in the vendor shop and conduct an effective and timely review and approval of documents submitted by the vendor. Also, expediters may be used in the vendor shop, depending on lead time for fabrication and/or complexity of equipment, to expedite a vendor's required documents submittal in a timely manner. Tracking and supporting of witness and hold points for selected inspection and testing activities are highly critical in moving fabrication forward—thereby completion of equipment in time for delivery. Also, photo records during inspection and packaging are encouraged to avoid questions of lost or missing equipment after the prolonged storage at a customer site.

Step 6: Preproduction

Preproduction requires attention to what needs to be completed prior to the start of fabrication. Once the fabrication starts, any further design changes may significantly impact cost and schedule of equipment delivery due to materials being used and cut for fabrication. Delay in critical equipment delivery may cause or contribute to a significant impact on the overall project cost and/or schedule. Examples of key elements for completion prior to preproduction are listed below.

1. All items listed for completion prior to fabrication start on inspection and testing list (item 13 of Step 1 "Technical and Purchase Specification") are completed.

2. All the documents identified to be submitted by the vendor on the documents submittal list (item 15 of Step 1 "The Technical and Purchase Specification") are required to be provided to the buyer for review and approval prior to the start of fabrication.

3. Vendor to ensure acquiring all the materials needed and be ready for fabrication.

4. All material tests and required certifications are obtained (e.g., ASME materials require certified material test report [CMTR] as part of the quality records).

5. Vendor to close all relevant audit findings from the past buyer's audits.

6. Buyer contract documents, including purchase specification, technical data sheets, drawings, lists, etc., must be updated to the latest agreed-upon revisions.

7. Relevant fabrication procedures approved by the buyer.

Step 7: Fabrication and In-Process Inspections

The fabrication and quality plans are to be submitted by the vendor as listed on the documents submittal list and approved by the buyer prior to start of fabrication. Fabrication progress to be tracked (using the example provided in table 9) to the completion of the fabrication. All in-process inspections required by the inspection and test plan must be followed as part of tracking the fabrication progress. Examples of in-process inspections are listed below.

1. Any new process during testing and fabrication.

2. Start of critical weld process and its qualification testing.

3. Equipment qualification testing.

4. Seismic qualification testing.

5. Critical steps or processes in the assembly of components.

Tracking of fabrication provides schedule status intelligence needed to take actions, if any, to expedite the remaining activities and processes in order to meet the final delivery date.

Table 9 – Fabrication Progress Tracking Example

Fabrication Activity No.	Activity Description	Completion Plan Date	Is Activity on QA Plan?	Completion Actual Date	Notes
From the vendor's fabrication/ quality plan	e.g., forging of valve body or testing a critical process	Date	Yes or No	Date	

Step 8: Testing and Final Inspections

Upon completion of fabrication, if component testing is required prior to assembly into a subsequent assembly and requires approval/review, it should be scheduled appropriately in the quality plan. Equipment will then be assembled for final testing and inspection. Final testing includes functional and performance testing and equipment qualification testing (for safety class equipment, as required for a nuclear application; other applications may have unique requirements). The testing must be completed to the acceptance criteria specified in the technical specification package of the contract.

Final inspection includes a review and acceptance of all the quality records and a quality records manual, with items listed on the quality records outline (see table 5), as supplied by the buyer and accepted by the vendor. The inspection includes the checking of final packaging and associated shipping documents stating the buyer's requirements.

Step 9: Release for Shipment

Packaging of the equipment for shipment should be completed using the approved procedures. Final inspection made by the customer or his designee during packaging and shipping would be useful later in dealing with the potential for claim of lost or damaged items from a shipment. Also, the customer representative can approve the release of shipment (see table 6 example). The storage procedures for any extended storage required must be reviewed and approved prior to equipment being delivered to the jobsite. A detailed packing list should be prepared not only for the top assembly in a box, but for all loose items such as nuts, bolts, software, or subassemblies. Prior to packaging of all the devices and parts in the boxes, photos should be taken for the records, which will be useful later.

The vendor then completes a quality certificate attesting to the completeness of the equipment and compliance with the procurement documents. The quality certificate also can be used to document release for shipment between the parties (see table 6 example) and the equipment's acceptance by the buyer or purchaser. Now the shipment is ready to ship for delivery to its predetermined destination, which is usually the jobsite.

Chapter 5
Phase III: Shipping, Delivery, and Receipt Inspection

This phase describes the processes for shipping, which includes transportation, delivery destination (including handling and storage at the delivery site), and receipt inspection by the site quality assurance team for final acceptance. The delivery also includes a spare parts list as required by the procurement documents (see table 10 example).

Step 10: Shipping

The shipping process requires early identification of a method of transportation (e.g., by air, sea, or ground) and selecting an agent to transport the equipment to its final destination with proper handling and care. Similar planning may be required for shipping hazardous materials and special software. These may require special permits, or various state or federal agencies may have special requirements for transporting these inside and outside the country.

For a typical equipment package, the following key steps are required for effective shipment and tracking.

1. For any overseas shipment, the buyer, transporter, or vendor (as defined in their respective contract) should review any items that may require a control permit for the country to export or import. Consult with the respective governmental agency and obtain the required permits for those items in the shipment. Example: paint containing critical chemicals.

2. Obtain weight, dimensions, number of boxes and their sizes, and planned ship date from the vendor. The ship date is from the vendor's shop to the delivery destination. See table 7 for a preliminary data/information request example. Preliminary information is useful in selecting a transporter or freight forwarder for the project. Final detailed information is required in preparation for shipment.

3. Notification to the applicable parties for final inspection and quality certification sign-off that includes a release for shipment (see table 6).

4. Send prenotification of a shipment to the transporter with information obtained in step 1 above.

5. Send final notification of the shipment with quality certification as obtained from step 3 and confirmation of final ship date.

6. Transporter to book a vessel, vehicle, or flight, as applicable, and notify the buyer and/or the vendor.

7. Transporter to obtain final packing list from the vendor and send a copy to the buyer.

8. Buyer to obtain quality record packages, and operational and maintenance (O&M) instruction manuals, and send to the equipment delivery destination (the jobsite).

The above steps do not include any administrative activities that may be required for transaction of payments and so on, since those may differ for individual companies. However, those can be added here by the user for completeness.

Step 11: Delivery and Receipt Inspection

Upon delivery of the equipment package, the equipment should be handled and stored according to the procedures provided by the vendor for safety and protection of the equipment and people. Proper tools shall be used, as specified in those procedures.

The quality team at the receiving end should check the receipt and O&M instruction manuals for complete quality. The quality team needs to check all the equipment, including any spare parts for a shipment, according to the packing list. The team needs to document any findings for the shipment, and notify the buyer and the vendor to address or resolve those findings or deficiencies.

Step 12: Final Acceptance

Final acceptance requires the vendor to resolve all the findings and/or remove those deficiencies to the buyer's or customer's satisfaction. Once all the findings are resolved, then final acceptance is provided and final payment is released.

However, in some cases, a small amount is retained through the contract until a specified number of operation hours of the installed equipment are reached. The retainer amount usually has a time limit, since many times the final installation and operation of equipment may be delayed significantly due to other factors of the project.

Table 10 – Typical Equipment Spare Parts List

Equipment Package: Number and Title:	HX-001 Heat Exchanger (Example)		
Vendor Name:	XYZ Co.		
Spare Parts	Part 1	Part 2	Part 3
Spare part number	J-1000-99		
Quality classification	S or G		
Spare part name/title or description	Shell to tube gasket		
Quantity	12		
Estimated price	$1500		
Delivery lead time	12 weeks		
Recommended operation and maintenance duration	3 years		
Design qualified life	5 years		
Shelf life	6 months		
Remarks			

Chapter 6
Summary and Applications

We have now covered all the phases of the procurement process and the steps involved in creating a complex equipment package, with examples cited for one of the major industries, a nuclear power plant-type project. Many elements can be simplified and used or applied effectively to engineered equipment procurement and its design considerations for many other industry applications, including small, medium, or large projects. Users can customize its use as applicable for an individual or a project need.

For cost and schedule effectiveness, engineers and designers should consider the following key factors, in addition to primary performance and intended functions.

—Equipment outline drawings showing envelop dimensions and orientation for ease with in-service inspection, general service, and maintenance.

—Control any design changes, including any material changes after the fabrication starts in vendor shop.

—Space allocation for installation and removal, as necessary.

—Perform feasibility of manufacturing engineered/designed equipment.

—Availability of potential vendors who can make it.

—Spare parts required during the life of the equipment and their availability.

In the quality assurance area, the following lessons learned from past experience are listed for consideration in future procurement.

—Go through inspection checklist before packaging.

—Plan to place an expediter at the critical supplier locations, especially for large and/or long lead equipment procurement.

—Take detailed digital photos of each side of package and of the equipment just prior to shipment.

—Proper tracking of receipt inspection findings and resolutions.

—For safety-related suppliers, review second-tier suppliers for pieces of equipment necessary for safety.

—Punch list of nonconformances per each package agreed to by the customer prior to quality certificate (QC) sign-off (see table 6).

—Assure that the in-process cleanliness practices and procedures provide adequate segregation or protection of stainless steel parts to prevent their contamination from the carbon steel fabrication processes. Assure that sufficient procedural controls are in place to prevent tools (grinding wheels, brushes, etc.) previously used on carbon steel from being used on stainless steel. Assure that the final cleaning process is adequate to remove all contaminants resulting from the fabrication process.

—The equipment supplier should consult with the paint manufacturer before applying any paint system for which the supplier has no experience base, to make sure it is appropriate for the application. Published technical bulletins do not necessarily address all variables that can affect paint integrity.

—Ensure all surfaces and edges, particularly welds, are smooth and clean prior to painting. Assure that the painted surfaces of components are thoroughly wiped down immediately prior to packing to remove all traces of surface contamination, such as carbon steel fragments and dirt.

—Assure that the shop preservation and packing procedure specifically addresses the type of preservative to apply to surfaces that potentially could be subjected to moisture condensation, sealing of all flanges, bolt holes, and other potential paths for intrusion of moisture, use of desiccant within confined air spaces, and ways of minimizing the entrapped air volume, such as shrink-wrapping, around exposed surfaces of components.

—Assure that care is taken during application of coatings and preservatives to make sure that complete coverage is achieved. Any creviced area where removal of the non-water-washable preservative would be difficult should be masked with an approved tape.

—The O&M manual should be one of the last items to be reviewed and approved, just before shipment of equipment, in order to include the critical test procedures. Most often those test procedures become the installation testing procedures (for validation of installation).

—The O&M manual should not include the packaging, shipping, and storage information. This information should all be in stand-alone documents. The storage document is required at the site prior to the equipment's arrival in order to prepare the adequate and compliant storage facility, unless equipment is ready to install upon arrival (e.g., just-in-time delivery).

In the commercial and contract area, the important lesson learned is not to sign a contract out of the procurement process (without technical specification and RFP) with premature design or inadequate technical specification. Otherwise, consequences are significant and can be critical to the overall project cost and delay of completion.

In one case, a premature contract for a large dollar amount (in millions) was signed with a vendor for one of the project's critical and complex subsystems having hardware and software elements. The design was not completed for that subsystem, and the final input (technical) data for the vendor were not available at the time of the contract. So, the required technical data inputs to the vendor flowed in a piecemeal fashion and continued changing due to evolving design. The design changes continued even after delivery of the subsystem to the jobsite. In the end, this premature contract cost more than double the original contract's value and caused significant schedule delays. This situation created a poor relationship between the parties and caused significant cost and schedule impact on the overall project. Therefore, to follow the established processes is vital for the procurement of all equipment and ultimately for the overall success of the project.

Finally, let me share a specific experience that I had twenty-five years ago, when I was leading a piping engineering service business for a nuclear services department. We signed a contract to deliver a large number of various pipe restraints for a nuclear power plant facility, based on a specific vendor's quotation. For those who may not know, pipe restraints are used to restrain a broken high-energy pipe in the event of a pipe break at a nuclear facility, in order to protect the surrounding safety-related equipment. We had a detailed design ready, with fabrication drawings and instructions for build-to-print for a qualified potential vendor. Today's plants do not require such a large number of pipe restraints due to advancements in design, technology, and materials.

Right after we signed the contract, the vendor that made the original quote went out of business and no longer made pipe restraint equipment. There was no other vendor available at the time, so our team searched for new potential vendors. However, we could not find any. We started looking for local area workshops and found one in our backyard that had the capability but no required quality program. The vendor was interested, so he hired a retired nuclear quality assurance expert and updated his shop to meet our quality requirements per NQA-1 standard. Afterward, we signed the contract. We completed the job with quality and delivered the equipment on time at less cost than the original vendor's quote. This successful experience was a reminder of the old saying, "Where there is a will, there is a way." We, our team, did have a will, and we found a way for success.

Abbreviations

ANSI	American National Standards Institute
API	American Petroleum Institute
ASME	American Society of Mechanical Engineers
ASTM	American Society of Testing and Materials
CFR	Code of Federal Regulations (US Regulations)
CMTR	Certified Material Test Report
EQ	Equipment/Environmental Qualification, as applied to
IEEE	equipment
	Institute of Electrical and Electronics Engineers
ISO	International Organization for Standardization
NDT	Nondestructive Testing
NFPA	National Fire Protection Association
NQA	Nuclear Quality Assurance, standard issued by ASME/ ANSI
O&M	Operation and Maintenance
QA	Quality Assurance
QC	Quality Certificate, as applied to equipment
RFP	Request for Proposal/Bid
RG	Regulatory Guide, as issued by USNRC
S, G	Quality groups; S for safety equipment and G for generic or non-safety equipment
USNRC	United States Nuclear Regulatory Commission

About the Author

Ramesh (RD) Patel was born in India and earned a bachelor's degree in mechanical engineering in 1967. He then moved to the United States for further studies and earned a master's degree in mechanical engineering in 1970. Since then he has worked in the nuclear power plant industry for forty years, including thirty-five with GE Nuclear Energy in San Jose, California, in various engineering capacities. Patel was a member of several nuclear-related American Society of Mechanical Engineers (ASME) code committees for over twenty-five years and is currently a contributing member of one of the nuclear piping code committees. He has coauthored five technical papers in the nuclear field and was a coinventor for three patents while working at GE Nuclear Energy. During his time in the nuclear field, Patel was involved in the development of critical procurement processes for the designing and purchasing of engineered equipment for a large project. The author used his forty years of experience and know-how to write this guide for the benefit of people involved in the procurement of engineered equipment in industrial/commercial projects, including engineers, designers, trainers, quality assurance engineers, and vendors.

A professional engineer in the state of California, Patel has been a consultant for the nuclear industry since retiring from GE in November 2009. He currently lives with his family in California.